Pollinators

BEES

Emma Bassier

DiscoverRoo
An Imprint of Pop!
popbooksonline.com

abdobooks.com

Published by Pop!, a division of ABDO, PO Box 398166, Minneapolis, Minnesota 55439. Copyright © 2020 by POP, LLC. International copyrights reserved in all countries. No part of this book may be reproduced in any form without written permission from the publisher. Pop!™ is a trademark and logo of POP, LLC.

Printed in the United States of America, North Mankato, Minnesota.

102019
012020

THIS BOOK CONTAINS RECYCLED MATERIALS

Cover Photo: iStockphoto
Interior Photos: iStockphoto, 1, 6, 7, 8–9, 11, 16 (top), 17 (top), 17 (bottom), 22, 27, 28–29; Shutterstock Images, 5, 16 (bottom), 19, 20, 21, 23, 25, 24; The Natural History Museum, London/Science Source, 12; USGS Bee Inventory and Monitoring Lab, 13; Dr Jeremy Burgess/Science Source, 14; Dennis Kunkel Microscopy/Science Source, 15

Editor: Connor Stratton
Series Designer: Jake Slavik

Library of Congress Control Number: 2019942666
Publisher's Cataloging-in-Publication Data
Names: Bassier, Emma, author.
Title: Bees / by Emma Bassier.
Description: Minneapolis, Minnesota : Pop!, 2020 | Series: Pollinators | Includes online resources and index.
Identifiers: ISBN 9781532165931 (lib. bdg.) | ISBN 9781532167256 (ebook)
Subjects: LCSH: Pollinators--Juvenile literature. | Bees--Juvenile literature. | Bees--Behavior--Juvenile literature. | Pollination by insects--Juvenile literature. | Insects--Juvenile literature.
Classification: DDC 595.799--dc23

WELCOME TO
DiscoverRoo!

Pop open this book and you'll find QR codes loaded with information, so you can learn even more!

Scan this code* and others like it while you read, or visit the website below to make this book pop!

popbooksonline.com/bees

*Scanning QR codes requires a web-enabled smart device with a QR code reader app and a camera.

TABLE OF CONTENTS

CHAPTER 1
Fuzzy Fliers 4

CHAPTER 2
Bee Bodies. 10

CHAPTER 3
Habitats 18

CHAPTER 4
Saving Bees 26

Making Connections 30
Glossary. 31
Index 32
Online Resources 32

CHAPTER 1
FUZZY FLIERS

A field of flowers shines in the summer sun. A bee lands on a pink snapdragon. A puff of yellow dust coats the bee's body. This dust is **pollen**. The bee sticks

WATCH A VIDEO HERE!

The bright color patterns of snapdragons attract bees.

its tongue into the flower to eat. Then it

flies to another flower.

Bees visit flowers to drink their **nectar**. It gives bees energy. Some also collect pollen to feed their babies. Bees have tiny hairs all over their bodies. Pollen collects on these hairs when they land on flowers.

pollen

BEE POLLINATION

A bee drinks nectar from a flower. Pollen gets on the bee's body. The bee flies to another flower. Some of the first flower's pollen falls into the second flower.

Bees are one of the most important pollinators on Earth. A pollinator spreads pollen from one flower to another. Plants use this new pollen to create seeds. Then new plants grow from the

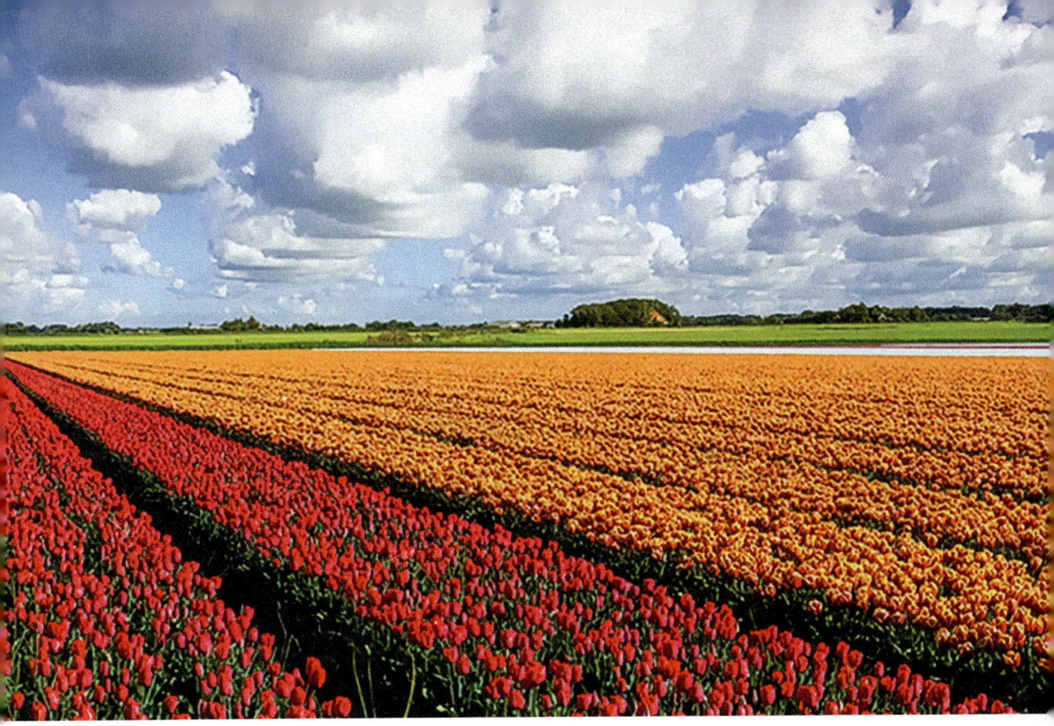

Plants with flowers make up approximately 90 percent of all plant species.

seeds. This process is why many plants need their pollen to spread.

DID YOU KNOW?

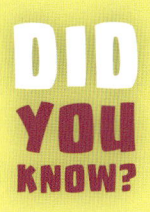

Honeybees can visit 5,000 flowers in one day.

CHAPTER 2
BEE BODIES

More than 20,000 types of bees live around the world. Many bees specialize in pollinating one type of flower. For example, squash bees tend to pollinate only the flowers of squash plants.

LEARN MORE HERE!

There are more than 2,000 species of sweat bee. Some are green!

Bees use their eyesight to find flowers. Like other insects, bees have **compound eyes**. They can see colors that humans cannot. Many flowers have these colors near their **nectar**. Bees can follow those colors to food.

Wallace's giant bee is the largest type of bee on Earth. This bee can grow up to 1.6 inches (40 mm) long.

The smallest types of bee can be fewer than 0.35 inches (10 mm) long.

DID YOU KNOW? Most types of bees hardly ever sting. Females' stingers are often short. And male bees do not have stingers.

In addition, some female bees have **pollen** baskets. These rows of stiff hairs line the bees' back legs. These hairs look like combs. They hold pollen. Bees can carry more pollen as they fly from flower to flower.

A pollen basket helps a bee collect more pollen.

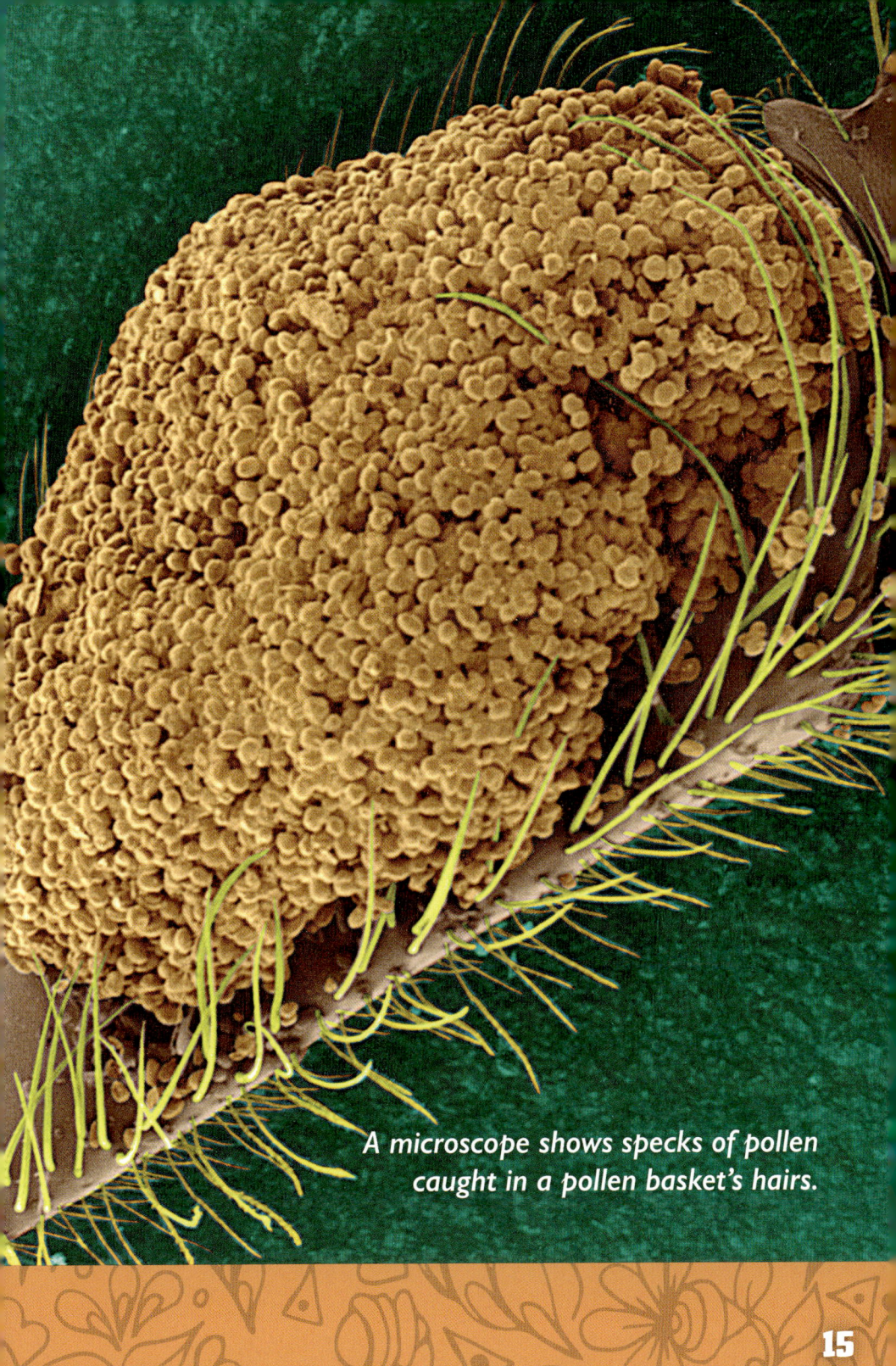

A microscope shows specks of pollen caught in a pollen basket's hairs.

LIFE CYCLE OF A BEE

A female bee lays eggs.

Larvae hatch from the eggs. Larvae are beginning forms of insects, often looking like worms.

Larvae grow and change into pupae. Pupae are the forms of insects between larvae and adults.

Pupae transform into adult bees.

Most adult bees live for a few weeks. A queen bee can live up to five years.

CHAPTER 3
HABITATS

Bees live in many countries around the world. Some bees are **native** to places with cold weather. Others are native to hot and rainy areas.

COMPLETE AN ACTIVITY HERE!

More than 4,000 types of bees are native to the United States.

One hive can be home to tens of thousands of bees.

Honeybees and bumblebees live in groups called colonies. They make hives out of wax. They turn **nectar** into honey in these hives. Then the bees

A honeycomb is made of wax cells. Each cell has six sides.

LIFE IN A COLONY

In a colony, bees have different roles. One female bee is the queen. She lays eggs. She can lay up to 2,000 eggs per day. Other female bees in the hive are workers. These bees build honeycombs. They keep the hive clean. They also gather food and feed the queen. Each colony has thousands of workers. All male bees are drones. They fly off to find a queen and start a new hive.

store the honey in **honeycombs**. Colonies use this honey for food.

In contrast, solitary bees live alone. They find shelter in places such as holes in trees or walls. Many solitary bees also dig homes in the ground. Solitary bees survive by making and eating beebread. Beebread is a mix of **pollen** and nectar.

Solitary bees often dig homes in dry or sandy soil.

Bee larvae eat beebread in a hive made from mud.

A bumblebee visits the flower of a tomato plant.

The types of flowers that bees visit depend on where they live. For instance, honeybees and solitary bees pollinate

DID YOU KNOW? Bumblebees are good tomato and pepper pollinators. Their buzzing shakes pollen loose from the plants' flowers.

many kinds of crops. Solitary bees also pollinate wildflowers, garden plants, and fruit trees. Blue orchard bees pollinate apple and plum trees.

Blue orchard bees pollinate in early spring.

CHAPTER 4
SAVING BEES

Some types of bees are **endangered**. Farm chemicals are one reason why. They help kill pests. But they often kill bees too. **Climate change** is another cause. Many bees can only survive in

LEARN MORE HERE!

In the United States, more than 1 billion pounds (454 million kg) of farm chemicals are used every year.

certain types of weather. When weather patterns change, a lot of bees die.

Many plants need bees to spread their **pollen**. If bees die, these plants will die too. Fewer crops would grow. People around the world would have less

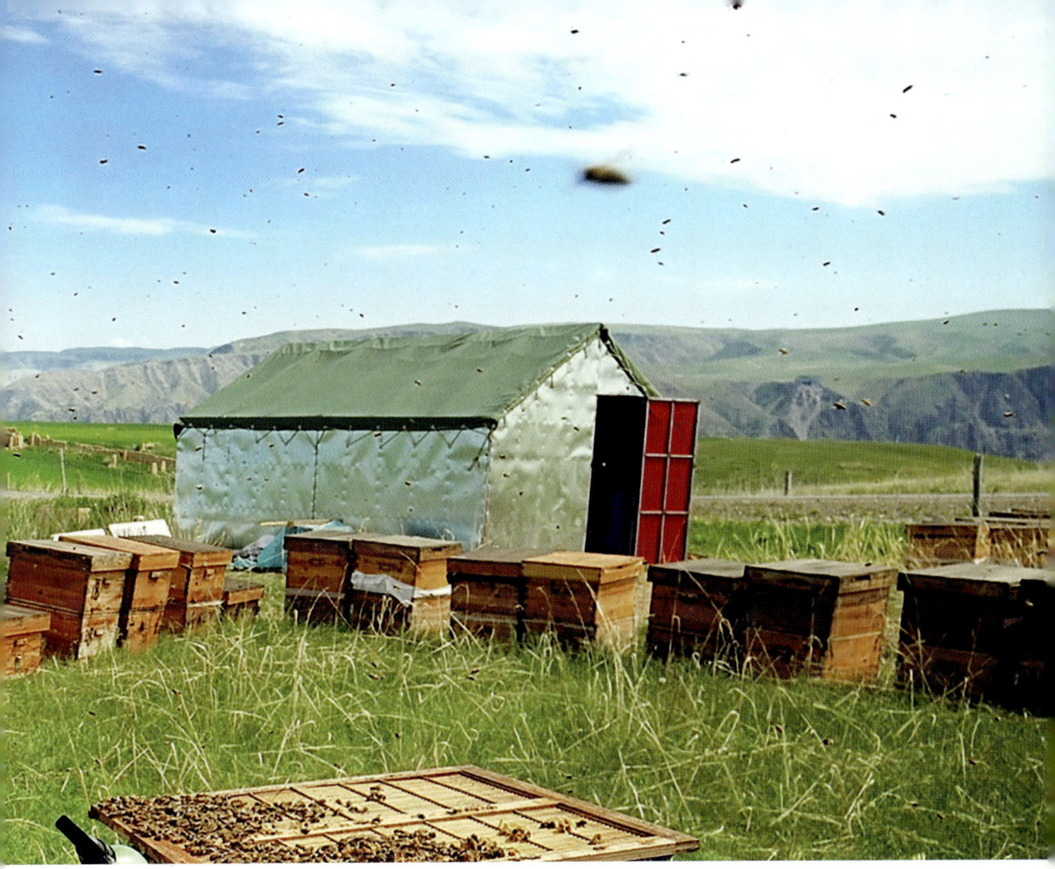

During the 2010s, beekeepers around the world lost large numbers of honeybees.

food. By saving bees, humans can help protect life around the world.

DID YOU KNOW?

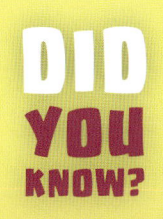

More than 33 percent of the world's crops rely on bees for pollination.

MAKING CONNECTIONS

TEXT-TO-SELF

Have you seen bees before? If so, what kinds? If not, where might you find them?

TEXT-TO-TEXT

Have you read books about other insects? What do they have in common with bees? How are they different?

TEXT-TO-WORLD

Bees are important pollinators. What are some ways bees help the world around them?

GLOSSARY

climate change – a crisis involving changes in Earth's temperature and weather patterns.

compound eye – a type of eye that is made of more than one seeing organ.

endangered – at risk of dying off completely.

honeycomb – a sheet of wax that holds larvae, pollen, and honey in a beehive.

native – naturally living in a certain area.

nectar – a sweet, sugary liquid made by a plant.

pollen – fine, dust-like stuff that flowers create and use to reproduce.

INDEX

beebread, 22

bumblebees, 20, 24

climate change, 26

colonies, 20–21

compound eyes, 12

drones, 21

honeybees, 9, 20, 24

larvae, 16–17

pollen baskets, 14

pupae, 17

queen bees, 17, 21

solitary bees, 22, 24–25

ONLINE RESOURCES

popbooksonline.com

Scan this code* and others like it while you read, or visit the website below to make this book pop!

popbooksonline.com/bees

*Scanning QR codes requires a web-enabled smart device with a QR code reader app and a camera.